What Is GEOLOGY?

Rebecca Woodbury, Ph.D., M.Ed.

Gravitas Publications Inc.

What Is GEOLOGY?

Illustrations: Janet Moneymaker

What Is Geology?
ISBN 978-1-950415-29-8

Published by Gravitas Publications Inc.
Imprint: Real Science-4-Kids
www.gravitaspublications.com
www.realscience4kids.com

Photo credits: Cover & Title Pg, Pornpimon, AdobeStock; Above, Mark Wilson, Public Domain; P.5. Beat Wormstetter from Pixabay; P.13. Pornpimon, AdobeStock; P.15. Mark Wilson, Public Domain; P.19. Alvise, AdobeStock

Do you ever pick up rocks and wonder how they were made?

I WONDER HOW THIS ROCK WAS MADE.

Do you sometimes look at mountains and wonder how they were formed?

I wonder if mountains were formed for hiking.

Have you ever wondered what is at the bottom of the ocean?

WOW! THERE ARE
MOUNTAINS
AT THE BOTTOM
OF THE OCEAN!

All of these questions are questions about **geology.**

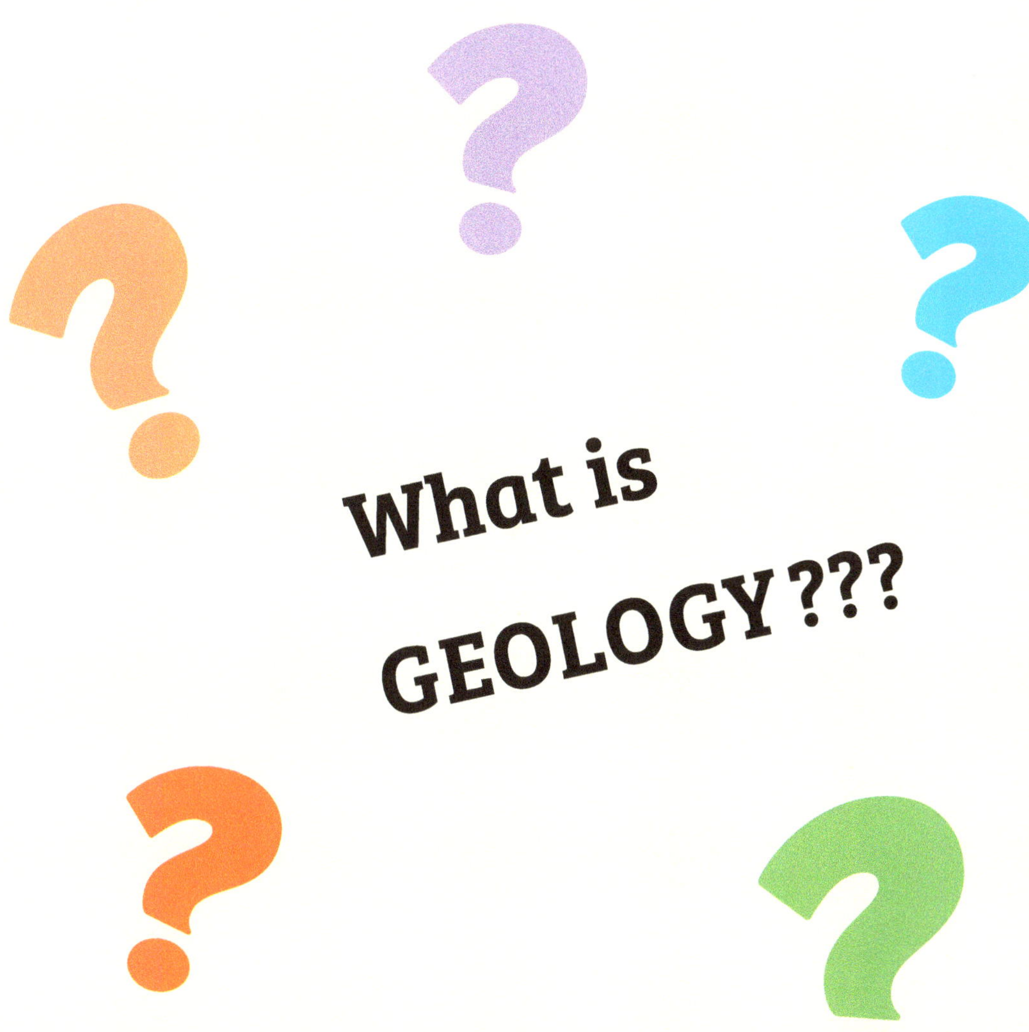

What is GEOLOGY???

Geology is the study of Earth.

GEOLOGY: Word Roots

geo- means Earth

-logy means study

GEOLOGY means:
the "study of Earth"

See how that word is put together?

Interesting!

A **geologist** is a scientist who studies Earth.

Geologists study changes in Earth over time.

Some changes take a
very long time to occur.

I WONDER HOW LONG IT TOOK FOR THE MOUNTAINS AND RIVERS TO GET HERE.

And some changes
happen more quickly.

We can learn about Earth by studying volcanoes!

By studying how Earth changes, we can learn how to take care of the planet where we live.

LET'S DISCOVER HOW TO KEEP THE AIR AND WATER CLEAN.
YES!
We LOVE Earth!

How to say science words

geologist (jee-AH-luh-jist)

geology (jee-AH-luh-jee)

planet (PLAA-nuht)

science (SIY-uhns)

scientist (SIY-uhn-tist)

study (STUH-dee)

volcano (vahl-KAY-noh)

www.ingramcontent.com/pod-product-compliance
Lightning Source LLC
LaVergne TN
LVHW060635110826
845147LV00014B/910
9781950415298